Educators Love STEP-UP Books.
So Do Children.*

In this exciting series:
- THE WORDS ARE HARDER (but not too hard)
- THERE'S A LOT MORE TEXT (but it's in bigger print)
- THERE ARE PLENTY OF ILLUSTRATIONS (but they are not just picture books)
- And the subject matter has been carefully chosen to appeal to young readers who want to find out about the world for themselves. These informative and lively books are just the answer.

*"STEP-UP BOOKS

... fill a need for precise informational material written in a simple readable form which children can and will enjoy. More please!"—EVELYN L. HEADLEY, *Director of Elementary Education, Edison, New Jersey.*
"I love them."—STEVE MEYER, *second grade pupil, Chicago, Illinois.*

The Story of Flight is the history of flying from ancient Greek fable to the exploration of space. It tells about the colorful era of ballooning, and the development of flying machines from the Wright brothers' airplane to the giant spaceships of today. "Space Age" children will find these adventures of bold, daring men and their marvelous machines fascinating reading.

The Story of *Flight*

By MARY LEE SETTLE
Illustrated by GEORGE EVANS

Step-Up Books   Random House
New York

To Jane Erskine with love

MARY LEE SETTLE was born in Charleston, South Carolina, and was graduated from Sweetbriar College. During World War II she served with the Women's Auxiliary Air Forces in England, and her interest in airplanes dates from this time. The most recent of her eight novels is the critically acclaimed *All the Brave Promises*.

Miss Settle is an assistant professor of English at Bard College, and lives in Barrytown, New York.

GEORGE EVANS spent a lot of time around airplanes when he was an aircraft mechanic in the Air Force during the Second World War. His skill as an artist combined with his knowledge of planes and aviation history result in marvelously detailed and exciting illustrations.

Mr. Evans studied at the Art Students' League in New York, and now lives in Levittown, New York, with his wife, two daughters, and a very smart dog.

© Copyright, 1967, by Random House, Inc. All rights reserved under International and Pan American Copyright Conventions. Published in New York by Random House, Inc., and simultaneously in Toronto, Canada, by Random House of Canada Limited. Library of Congress Catalog Card Number: 67-24456. Manufactured in the U.S.A.

Contents

1	Dreams of Flight	3
2	The First Leap	7
3	The Montgolfier Brothers	13
4	Everyone's Up in the Air	17
5	The Wright Brothers	25
6	The First Airplane	29
7	Dare-Devils	35
8	War in the Sky	40
9	The Red Baron	47
10	New Adventures	51
11	The Spirit of Saint Louis	54
12	Farther and Faster	61
13	The Battle of Britain	65
14	The Flying Bomb	69
15	Rockets	73
16	Man in Space	79
17	The Space Race	83

1
DREAMS OF FLIGHT

For ages and ages men watched the birds wheel and fly through the air. It took men many days to walk through the woods. But the birds could fly quickly over the trees. Men could not see beyond the next hill. But the birds could look down on the hills and fields.

People dreamed of flying. They began to make up stories about men who could fly.

One story was told around 3,000 years ago. It was about a man named Daedalus. He lived on the island of Crete, near Greece. He worked for the king of the island, King Minos.

Daedalus did not like Crete. He wanted to go to Greece. But King Minos would not let him go. So Daedalus decided to try flying away across the sea. He made wings of wax and bird feathers. He made wings for his son, Icarus, too. He told Icarus not to fly too close to the sun or too close to the water.

Daedalus took off across the sea. His son followed him. For a long time they flew on.

Then Icarus began to look around him. He could see the islands under him. They were so beautiful that he wanted to see more. He began to fly higher and higher. His father called to him to come back. But he would not listen. He flew nearer and nearer the sun. The wax in his wings melted. Icarus fell into the sea and was drowned.

Daedalus flew on to Greece alone.

2
THE FIRST LEAP

The story of Daedalus was only a story. But men really did think they could find a way to fly. For hundreds of years they tried. They made kites. But no one could make a kite big enough to hold a man in the air.

In Italy men began to make wings of feathers and cloth. The men put them on their backs. They climbed up high towers and jumped off. They were called tower jumpers.

Most of them were killed.

But in the year 1490 a man named Danti climbed a tower in Perugia, Italy. Far below many people stood watching. Danti jumped!

The wind caught his wings. For a few seconds he glided. Then his left wing broke. He crashed to the ground. He was badly hurt. But he had made the first wings that had worked at all.

In 1507 a man named John Damian made wings of feathers. He wanted to fly from the top of a castle in Scotland. He climbed up and jumped. He fell right to the ground and broke his leg. He said it was because he had used hen feathers. He said hens were not used to flying!

In China a man thought of using toy rockets to get himself up into the air. They were like the sky rockets we see on the Fourth of July. They made showers of pretty sparks.

The man was named Wan Ho. He had 47 little rockets tied to the back of a chair. He held two kites in his hands to steer him. He had 47 servants light the 47 rockets all at the same time.

There was an explosion.

Nobody ever saw Wan Ho again. After that some people said he was the man in the moon.

About the same time, in Italy, a man was working on new ideas.

His name was Leonardo da Vinci. Leonardo studied the way birds flew. He studied the air itself. He made flying models that rose all by themselves from the floor of his workroom. These models went straight up just the way helicopters do today.

But they were far too little to carry a man into the air.

Leonardo watched the way leaves fell through the air to the ground. This gave him an idea. He made drawings of the first parachute.

When Leonardo died he left many drawings of his flying models. He drew them carrying men. He was sure men could really fly. But he never made a real parachute, or a helicopter big enough to lift a man into the air.

Leonardo died in the year 1519. Almost 300 years later men were still trying to fly. They made many different kinds of wings. They made parachutes. They even tried to make flying boats. But nothing seemed to work.

3
THE MONTGOLFIER BROTHERS

In 1781 a new discovery was made. In a little town in France lived two brothers. Their names were Joseph and Etienne Montgolfier. They were paper makers.

Joseph often watched bits of wood and paper rise up from the heat in the fireplace. He wondered if a bag of hot air would rise, too.

The brothers began to play a very dangerous game. They made fires under paper bags. The heat made the bags rise into the air.

On June 5, 1783, the two brothers set up a huge balloon in the marketplace of the town. There was a pit under it filled with straw and wool.

A crowd gathered to watch. A fire was started in the pit. The balloon filled with hot air.

The balloon rose up into the sky and floated away. It floated through the air for more than a mile and landed in a field.

Other balloons were built. One came down on a farm. The farmers thought it was a monster. They ran for axes and pitchforks. They cut the balloon to pieces.

Joseph and Etienne soon built a bigger balloon. They put beautiful drawings on it. And this time they put a basket underneath it. They put a duck, a sheep, and a rooster in the basket. The three animals took off into the sky. Not far away they landed safely.

The Montgolfiers had shown that a balloon could carry living things through the air. They decided to try to make one big enough to carry people.

4
EVERYONE'S UP IN THE AIR

On November 21, 1783, people in the city of Paris, France, gathered on the rooftops. They looked out of windows. They waved handkerchiefs.

In the air were two men in a Montgolfier balloon. The pilot was named Pilatre de Rozier. He and his friend rode over the city for 25 minutes. Sometimes they floated down close to the rooftops. Sometimes they floated high over the city. They were the first people to fly in the air.

On January 7, 1785, a man named Pierre Blanchard left England in a huge, new balloon. He was with an American friend, Doctor Jeffries. They were trying to cross the sea between England and France.

On the way across, the balloon began to sink closer and closer to the water. The two men threw everything they could into the sea to make the balloon lighter.

They threw anchors and ropes. They threw sandwiches. They threw cheese and wine. They even threw their pants. Then, just as they were going to fall into the sea, a wind lifted the balloon up. They went on and landed safely in France. They were the first men to go from one country to another by air.

Men made lots of balloons filled with hydrogen gas. Hydrogen gas is lighter than air. It rises up into the air. This gas made the balloons fly higher. And it made them stay up in the air longer.

Balloon flights became a sport. Everybody wanted to fly.

In England a plump actress, Mrs. Letitia Anne Sage, wanted to make a balloon flight. She took a friend named Mr. Biggin with her. The morning they were to go up lots of people came to see the famous lady fly. She wore her best dress. She wore a big hat with feathers. The pilot put on a bright red uniform. They all climbed into the balloon.

Mrs. Sage waved to the people. The people waved back. The pilot let go the ropes.

The balloon did not move. Mrs. Sage was too fat.

The pilot was too polite to say so. He got out. Up went the balloon!

It finally landed on a farm. Mrs. Sage and her friend began to walk across a field. They were pleased. But the farmer was not. Fat Mrs. Sage was trampling on his beans.

In America there was a famous balloonist named Thaddeus Lowe. In 1859 he made a huge balloon. He wanted to fly across the Atlantic Ocean. But before he could try, the balloon burst.

Thaddeus Lowe never had another chance to fly across the ocean. War came before he could try again.

The war was called the Civil War. It was between the states of the North and the states of the South.

Lowe began to make balloons for the army of the North. In 1862 he had seven balloons. They could fly high over the battlefields. The men in them could see more than any man on the ground. They could see the armies getting ready to attack. There were wires going from the balloons to the ground. The men sent telegraph messages down them.

On May 31, 1862, there was a big battle at Fair Oaks, Virginia.

Lowe went up in a balloon called the Intrepid. He looked down over the battlefield. He saw what the men at Northern headquarters could not see. The South was winning.

Lowe quickly sent a message to the men at headquarters. They sent out more Northern soldiers. Soon the Southern soldiers fell back.

People said that if Lowe had not sent his message, the South would have won a great battle.

The Southerners wanted a balloon, too. But they did not have enough money to buy one. So the ladies gave their pretty silk dresses. The balloon made from them looked like a big patchwork quilt.

5
THE WRIGHT BROTHERS

One day in 1878 a minister in Ohio brought his two sons a toy helicopter. The little boys' names were Wilbur and Orville Wright. Mr. Wright threw the helicopter into the air. It flew right up to the ceiling.

The boys were very excited. They loved flying kites. But here was a toy that did not need wind to make it rise.

The Wright brothers were good at making things. As they grew older, they made a printing press. They made bicycles. And they made lots of kites. They flew the kites and studied their flight.

Then they read about something new called a glider. It was made and flown by a man named Otto Lilienthal.

Lilienthal would strap the glider to his back. Then he would run fast down a hill. The glider would be lifted and held up by the wind.

Lilienthal made many flights. Then one day a strong wind caught his glider. Lilienthal lost control. He crashed to the ground and was killed.

The Wright brothers learned from Otto Lilienthal. They made gliders and learned to fly them. They flew many gliders from a place called Kill Devil Hill, near Kitty Hawk, North Carolina. The winds there were good for flying. And there was smooth sand for landing.

The brothers took turns flying the glider. They put bags of sand on it. They wanted to see how much weight it could carry. They hoped it could carry a motor.

6
THE FIRST AIRPLANE

On the morning of December 17, 1903, the wind blew cold over Kill Devil Hill. The two Wright brothers had made a new flying machine. It was like a glider with two pairs of wings. It was made of cloth and wood and wire. There was a motor in it. It had two propellers. It was an airplane.

The airplane was resting on a little track.

Orville climbed onto the airplane and lay down. Wilbur stood ready to run beside it. He had to hold it onto the track. They started the motor. The airplane began to shake all over. Then it moved slowly down the track.

Suddenly Wilbur felt the wings lift away from his hands. Orville was in the air! He was flying!

The airplane stayed in the air for only 12 seconds. It landed with a bump. But Orville Wright was the first man to fly an airplane.

The Wright brothers set to work making their airplane better. They made longer and longer flights.

They worked for four long years.

Then, in 1908, Wilbur sailed to Europe. He took an airplane on the ship with him.

In France he flew his plane for hundreds of people. He became very famous. Even kings came from far away to watch him fly.

All over Europe men saw that the Wright brothers had made a very great discovery. They saw many uses for the airplane. They knew that the airplane would be very important in war.

That same year, in America, men in the Army asked to see the new planes fly. They wanted to see for themselves if they really worked.

A big crowd went to Fort Myer, Virginia. Orville took off and flew for an hour. When he landed, the crowd cheered and cheered.

The Wright brothers had shown the world that they could make and fly an airplane.

After the Wright brothers, other men built and flew airplanes. The airplanes were all shapes and sizes. There were no airports. The pilots landed in any field they could find. Sometimes they had to walk a long way home.

7
DARE-DEVILS

Early in the morning on July 25, 1909, Louis Bleriot was in a cow field in France. He sat in a little flying machine. Men held on to the wings to keep the airplane steady. The motor was started.

Bleriot raised his hand. The men let go. The little airplane flew up over the beach. Ahead of Bleriot was the English Channel. He was trying to fly all the way to England.

There were no fields to land in. Only miles of water lay ahead. He flew out over the sea. He flew close to the water.

At last he saw the white cliffs of Dover, England. He was trying to land at Dover Castle. But he had been blown too far east by the wind. He turned into the wind and flew along the coast. Down in the sea he could see battleships.

Then he saw a big flag spread on the ground to show him where to land. The wind was so strong he almost crashed when he landed.

Bleriot stepped out of his airplane. It was still early in the morning. No one was there but one policeman.

But soon the news spread that he had landed. Crowds of people came. Some were campers. One of the campers was so excited that he forgot to put his clothes on.

Now there is a stone monument where Bleriot landed. He was the first man to fly an airplane over the open sea.

Flying now became a great sport. Soon there were many air races. The pilots were called Dare-Devils.

On April 27, 1910, there was a great airplane race in England. It was between an Englishman named Grahame-White and a Frenchman named Louis Paulhan. They were going to race 175 miles from London to Manchester.

They both took off in bad weather. At first Paulhan was way ahead. Night came. He landed his plane. He went to bed and settled down to sleep until dawn.

Grahame-White knew he could not wait until dawn if he wanted to win the race.

No one had ever flown at night. But Grahame-White was sure he could fly by the light of the moon. He took off.

Paulhan waited for the first light of day. Then he took off. He caught up with Grahame-White.

For a while they raced neck and neck. Grahame-White ran into a strong wind. His airplane's engine stopped. He had to land in a field.

Paulhan hit strong winds, too. But he kept going until he got to Manchester.

When Paulhan landed, there were thousands of people waiting to see him. He was tired. But he had won the race.

8
WAR IN THE SKY

In 1900 the first giant air ship had been made by a German named Count Zeppelin. It was filled with hydrogen like a balloon. It looked like a huge sausage. It carried lots of people through the air.

In 1914 a war began in Europe.

England, France, Russia, and Italy were on one side. They were called the Allies. They were at war with Germany and Austria. It was the First World War.

Germany began to send Zeppelin airships over Allied cities. They dropped bombs on them. For the first time people saw that cities could be hit from the skies.

At first airplanes were not used for fighting. The pilots just flew over the enemy. Then they came back and told what they had seen.

Early in the war, pilots on both sides waved as they passed. Then they began to throw bricks. Soon they were shooting at each other with pistols, rifles, and shotguns.

Pilots learned to wheel and turn. They learned to fly their airplanes near the enemy. The gunners in the back fired out the side.

Then a Dutchman named Anthony Fokker made a new airplane for the Germans. It was named after him. The Fokker carried a new kind of machine gun on its nose.

The gun fired in between turns of the propeller. Now the pilot could fire straight ahead. He could fly without a gunner. The Fokker was like a flying machine gun.

A new kind of air fighting began. These air fights were called "dogfights."

The pilots knew an enemy by the colors on his airplane. The German planes had black crosses painted on the wings. The English planes had circles of red, white, and blue on their wings.

Soon there was a new kind of war hero. He was a pilot who had shot down at least five airplanes. He was called an "ace."

One of the first aces was named Oswald Boelcke. He was a German. He shot down 40 Allied planes. He was so good even his enemies liked him. He was killed when one of his own men flew into him.

At Boelcke's funeral Allied pilots flew over and dropped wreaths.

The greatest English ace was a major named Edward Mannock. He shot down 73 enemy planes.

America was not in the war. But some American pilots flew with the French air force. They were called the Lafayette Escadrille. They used a new plane called the Nieuport 17. The Nieuport's motor was greased with castor oil.

The men in the Escadrille were wild and brave. They all loved to fly. They kept two pet lions called "Whiskey" and "Soda."

In 1917 the United States joined the Allies in the war. The men of the Lafayette Escadrille joined the other American pilots.

9
THE RED BARON

The greatest ace in the war was a German. He was a handsome, brave man named Baron von Richthofen.

Richthofen was the leader of the best group of pilots in the war. The group was called the Flying Circus. Its airplanes had bright red noses and crazy markings. But Baron von Richthofen's airplane was all red.

The Allied pilots always knew him by his red plane. They called him the Red Baron.

The Red Baron attacked with the sun behind him. This made it very hard for his enemies to see him. The sun was always in their eyes.

Then on the morning of April 21, 1918, a patrol of five planes from the Flying Circus took off to look for Allied planes. They were led by von Richthofen. For the first time he did not worry about keeping the sun behind him.

About the same time an English patrol started out. It was led by a Canadian, Arthur Royal Brown.

The two patrols met in the air. The planes circled each other. They turned and wheeled trying to shoot each other.

Von Richthofen started after one of the English planes. He flew down behind it. He was in the best place to shoot it down.

But Captain Brown saw the bright red plane. He flew down behind von Richthofen. For once the Red Baron did not look behind him. Brown fired one burst from his machine gun. He saw von Richthofen look up surprised. The red plane went down and crashed in the English part of the battlefield.

When he died, the Red Baron had shot down 80 Allied planes.

10
NEW ADVENTURES

In November, 1918, the war was over. The Allies had won. Each side had worked to make better airplanes. Now airplanes were much faster and safer than before the war.

When peace came, men found new ways to use what they had learned. In America many old war planes were made into mail planes.

Soon men were flying planes with bags of mail from city to city. They flew all over the country. They flew over mountains and across deserts.

It was hard and lonely work. The pilots had to fly all night. There was no one to show them the way. There was no one to tell them what the weather would be. They found out about rain, wind, and snow as they flew. Sometimes it took two weeks to fly across the country.

At each stop a pilot would write down a report of his flight. One report was very short. It said, "Flying low, only place available to land, on cow, killed cow, scared me, Smith."

In 1919 two men named Whitten Brown and John Alcock decided to try to fly over part of the Atlantic Ocean.

On June 14 they took off from Newfoundland. Newfoundland is an island. It is just off the coast of Canada.

The two men took turns flying the plane. They flew out over the ocean where there was no place to land. They flew over icebergs. They flew into rain and fog. They flew for 15 hours and 57 minutes.

At last Alcock and Brown landed in a swamp in Ireland. They were bruised and muddy. But they were not hurt.

They were the first men to fly so far across the open sea.

11
THE SPIRIT OF SAINT LOUIS

It was May 20, 1927. It was very early morning. It was raining. A small plane waited at an airfield in New York. It was called the Spirit of Saint Louis. A tall young mail pilot named Charles A. Lindbergh was looking over his plane. He was going to try to fly all the way across the Atlantic Ocean to France. And he was going to do it alone.

Lindbergh put a bottle of water and a bag of sandwiches into the Spirit of Saint Louis. He got into the airplane. It started down the field. It began to climb slowly. It cleared the telegraph wires at the end of the field by only 20 feet.

Lindbergh flew north up the coast of America. He flew close over the ocean waves. He turned east. At last he flew into the sunlight.

Then he ran into rain. It beat on the cloth wings of the little plane. Below him were big chunks of ice. He flew on all day. He flew into fog. He flew into night. There was nothing but clouds and darkness around him.

He saw a storm ahead. Snow began to blow around his little plane. He shined his flashlight on the wings. There was ice on them. The storm was throwing the plane around in the air.

Then he saw a few stars. The storm was over. The moon came out. It lit up the ocean under him. It lit the clouds around him.

He was all alone in the sky in the middle of the Atlantic Ocean.

Lindbergh fought to keep awake. He saw the white light of morning ahead. He had been flying a day and a night. He was 1,800 miles from New York. He was 1,800 miles from France. He was lonely and frightened.

He flew on and on. Then he saw a gull. That meant land was near!

He saw fishing boats. He flew his plane down close to one. He yelled out, "Which way is Ireland?" The fishermen were too surprised to say a word.

Then, there was Ireland below. He flew over the Irish Sea. He flew over a part of England. He flew all the next day. It was night again.

Down below him were the lights of Paris. He wondered if anyone would be at the French airfield. He saw landing lights on the airfield. They were laid out in two long lines so he could land between them. He had been flying alone for 33 hours. He brought his airplane down and landed.

He was the first man to fly across the whole of the Atlantic Ocean. And he had done it all alone.

He stepped out of the airplane. People were running toward him in the darkness. He was so tired he could not believe his eyes.

There was a huge crowd waiting at the airfield to meet him. People lifted him up on their shoulders. He had been so alone. He did not know people cared so much about his flight across the ocean. He was a shy, brave hero.

That night he slept in Paris. He had to ask for pajamas. There had been no room for his own in the Spirit of Saint Louis.

12
FARTHER AND FASTER

All over the world men found new uses for airplanes. Pilots took mail and food to people who were snowed in. They picked up sick people and took them to doctors. They flew over jungles. They flew over mountains. They brought help to people who lived far from towns and cities. These pilots were called bush pilots.

Now that airplanes were used for so many things, men worked hard to make them better and stronger. They were made of metal instead of cloth. They flew faster and faster. Passenger travel began. More and more people flew. A plane was much quicker than a car or a train.

In 1928 Kingsford Smith crossed the Pacific Ocean. In 1931 Wiley Post flew all the way around the world. It took a little over eight days. Great air races were held. Fast planes were made to race in them. Many different countries sent their best planes.

One race was called the Schneider Cup Trophy.

In 1931 the Schneider Cup Trophy was won by the fastest plane yet. It flew at 407 miles an hour.

An Englishman named Reginald Mitchell made the plane. Mitchell was sure another war was coming. He was afraid Germany was going to attack England. He knew England would need fast fighter planes.

Most people did not believe there would be a war. But Mitchell went right on making better planes.

In 1935 a new plane took to the air. Mitchell called it the Spitfire. The Spitfire was the fastest and most beautiful plane anyone had ever seen.

13
THE BATTLE OF BRITAIN

In 1939 war did start in Europe. It was called the Second World War. England, Russia, and France were on one side. Germany and Italy were on the other.

For years Germany had been at work building a mighty air force. It was now the best in the world. The Germans had fast fighters called Messerschmitts. And they had huge bombers.

Every night English fighter pilots took to the air in their Spitfires to fight the Germans.

September 12, 1940, was the worst night of the German bombing. The bombers dropped thousands of fire bombs on London. They wanted to set the whole city on fire.

Many parts of London burned. The fires could be seen for miles. All night thousands of men worked to put out the fires. They saved many people from their burning homes. Other people stayed in the subways under the ground.

Up into the night went the English Spitfires. High above the city the battle raged. There were many more German planes than English. But the brave English pilots fought on through the long, terrible night.

Morning came. London was still burning. But the German planes had been beaten back. Never again did Germany send so many planes over England.

The great air battle over England was called the Battle of Britain.

Winston Churchill was the head of the English government. After the battle he said, "Never in the field of human conflict was so much owed by so many to so few."

Churchill was saying that England had been saved that night. And it was the few brave English pilots in their Spitfires who had done it.

14
THE FLYING BOMB

On Sunday morning, December 7, 1941, airplanes took off from the decks of warships in the Pacific Ocean. The ships and airplanes belonged to Japan. They flew over the Hawaiian Islands. They bombed the American navy at Pearl Harbor. It was the first time any part of America had been bombed.

Japan had joined the war. It was on the German side. The United States went to war against them.

All this time both the English and the Germans were working on a new kind of airplane. It did not have a propeller. Its engine sucked in air from the front. The air mixed with fuel inside. It blasted out the back. This pushed the plane through the air. It went faster than any plane had ever gone. It was called a jet.

England's jet plane was called the Gloster Meteor.

Then one night over England, a new sound was heard in the air. It sounded like a motorcycle.

People thought it was an airplane. It had wings. It had a jet engine. But it did not fly like an airplane. It did not turn or twist.

The thing did not have a pilot. It was flying by itself. It was a flying bomb!

Germany called the flying bomb a V1. Thousands of them were sent over England. Some of the English pilots flew up in Gloster Meteors. They fired at the bombs with their machine guns. Many of the flying bombs blew up before they hit the ground.

15
ROCKETS

One day in 1944 the Germans fired a terrible new weapon. They called it the V2.

The V2 had no wings. It shot up into the air for 60 miles. It moved at 3,600 miles an hour. It made a half circle and came back down to earth. When it hit the ground it exploded. It killed many people.

It was a huge rocket.

The V2 was not the first rocket used in war. The English had used them in a war against the United States in 1812. But those rockets were too small to do much harm.

For years men worked on making bigger rockets. Then in 1926, an American, Robert Goddard, made a new kind of rocket. It was bigger. It could fly farther. On March 28, 1935, he sent one 4,800 feet in the air. It went 550 miles an hour.

In Europe men studied Goddard's new rocket. They learned from him how to make better rockets. But no one had ever made a rocket as huge and powerful as the German V2. It weighed 28,000 pounds!

The V2 did not help the Germans win. By May, 1945, Germany had lost. The war was over in Europe.

V2 rockets were brought to the United States and Russia after the war. Men found that the rockets could be fired a long way toward space.

There is a layer of air around the earth 100 miles thick. It is called the atmosphere. Beyond it is space.

Space had seemed too far away for men to explore. But now they began to think rockets might reach there.

In February, 1949, a V2 rocket stood ready to be fired. It was out in the desert near White Sands, New Mexico.

A small rocket was attached to the nose of the V2. It was called a WAC Corporal.

The men who were going to fire the rocket went into a building. It was under the ground, away from the blast. The men listened to a loudspeaker. The countdown began. "10, 9, 8, 7, 6, 5, 4, 3, 2, 1, ZERO!" At the word "ZERO!" a firing button was pushed.

There was a great blast of fire. The rocket went straight up.

The V2 became a little spot in the sky. For 20 miles it went straight on up. By then, all of its fuel was burned up. Then the engines in the WAC Corporal started.

The WAC Corporal went on up on its own. It kept going for 250 miles. It went beyond the atmosphere.

The first rocket had been fired into space.

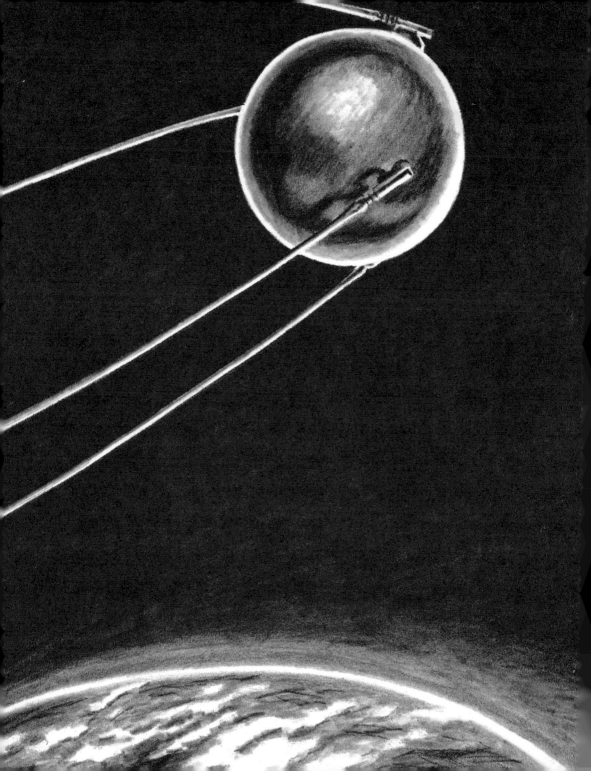

16
MAN IN SPACE

On October 4, 1957, the Russians fired a rocket with a small machine on top of it. The machine looked like a ball with spikes on it. It was called Sputnik 1.

In space the rocket burned out. The Sputnik kept on going. It went around and around the world.

A month later, the Russians sent up Sputnik 2. In it was a little dog named Laika. She went around the world six times. She landed safely. This proved that a live animal could go into space and come back. The time had come for a man to try it.

On April 12, 1961, a Russian pilot named Yuri Gagarin stood beside a new spaceship. It was called the Vostok 1. It weighed five tons. It was on top of a huge rocket.

Gagarin was wearing very heavy clothes. He could hardly move. He was helped into the Vostok. There was just room in it for one man. He was strapped into his seat.

The rocket was fired. Gagarin was carried into space.

In space, the burned-out rocket dropped away. Vostok 1 started to circle the world. Gagarin looked out into space. He could see the world. No one had ever before been far enough away to see its shape.

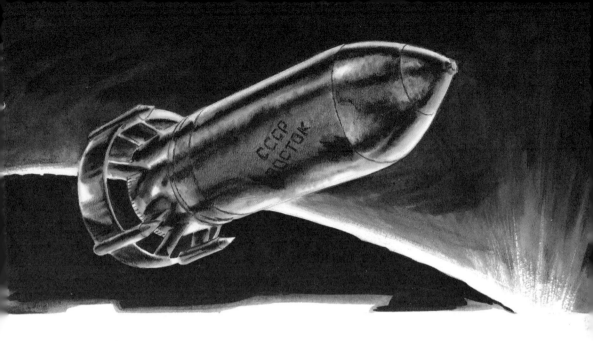

He could see the line between the atmosphere and the black of space. He said, "It is of a soft blue color and the entire change from blue to black is most smooth and beautiful."

Gagarin landed by parachute near a little Russian town. He had made the first trip of any man into space. He had gone around the world in one hour and 48 minutes.

17
THE SPACE RACE

Ever since the flight of Sputnik 1, the Americans and the Russians have been in a race into space. It is like those old races to see who could build the fastest airplanes.

On February 20, 1962, the first American went around the world in space. His name was John Glenn. The name of his spaceship was the Friendship 7.

He went around the world three times at five miles a second.

Since Sputnik 1, many uses have been found for space machines. The ones that go around the world are called satellites. Some of them send radio messages. Some send television pictures. One satellite sends television pictures between Europe and America.

There are satellites that can tell men about the weather. High in the atmosphere there are clouds. The satellites tell which way the clouds move. From this, men can tell more about what the weather will be.

Both Russia and the United States have sent machines deep into space. These machines send pictures and radio messages back to the earth.

They tell us what it is like on the moon and the far-away planets.

One space machine flew past the planet Venus. It sent back messages about it. One message said that it was hotter than boiling water.

It is amazing what men have done with their flying machines in such a short time.

The Wright brothers' helicopter was only a toy. Today helicopters are big machines that can carry many people and land on a rooftop.

Charles Lindbergh flew across the Atlantic Ocean alone. It took him 33 hours. Now a jet is being built that can carry 500 people. It can cross the Atlantic in five hours.

A tower jumper might not be too happy just sitting in an airplane with 500 other people. But there is one flying machine that would be sure to please him. It is a little rocket a man can strap on his back. He can turn it on and zoom around all by himself.

It took men thousands of years to learn to fly. Today they use the sky as easily as they use the sea.

Now men are sending machines into space. They are learning more and more about the stars and the planets. Men want to find out all they can about these places.

Some day they will be going there.

The STEP-UP Books

NATURE LIBRARY

ANIMALS DO THE STRANGEST THINGS
BIRDS DO THE STRANGEST THINGS
FISH DO THE STRANGEST THINGS
INSECTS DO THE STRANGEST THINGS

Story of AMERICA

Meet THE NORTH AMERICAN INDIANS	THE ADVENTURES OF LEWIS AND CLARK
Meet THE MEN WHO SAILED THE SEAS	Meet ANDREW JACKSON
Meet CHRISTOPHER COLUMBUS	Meet ABRAHAM LINCOLN
Meet THE PILGRIM FATHERS	Meet ROBERT E. LEE
Meet BENJAMIN FRANKLIN	Meet THEODORE ROOSEVELT
Meet GEORGE WASHINGTON	Meet JOHN F. KENNEDY
Meet THOMAS JEFFERSON	Meet MARTIN LUTHER KING, JR.

THE STORY OF FLIGHT

A Map
showing some
Important Places
talked about in
The Story of Flight